MO

Battery electr
if necessary.

WEEKLY CHECKS

1. Gearbox oil colour – oil must not be discoloured.

2. Water filter – if there's a lot of floating weed around, check daily.

DAILY ENGINE CHECKS

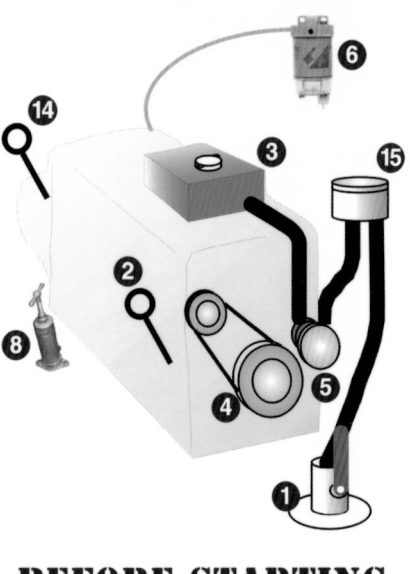

BEFORE STARTING

1. Cooling water cock open.

2. Engine oil level.

3. Fresh water coolant level.

4. Drive belt tension.

5. Leakage from the underside of the raw water (sea water) pump. If dripping (from the hole), replace pump internal water seal.

6. Water separating fuel filter – check for water in the bowl.

7. Any oil or water leaks.

8. Charge stern-gland with oil or grease if applicable – check for leaks – a couple of drips per minute is normal with this traditional type of gland.

No drips usually indicates that the gland is too tight and will overheat. But modern 'dripless glands' should not leak.

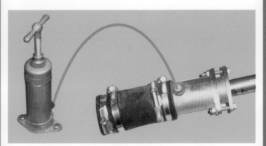

DURING CRUISE

- If possible have a quick look at the engine(s) hourly – don't get any clothing or gear caught in moving parts, this can be fatal. Basically you are looking for leaks and things working loose.
- Grease stern-gland if necessary.
- Always try and run your engine under load.

AFTER STOPPING

- Isolate engine start battery if this needs to be done manually.
- Switch off unnecessary electrical loads
- Log engine running time.

ENGINE WON'T TURN

1. Is battery switch 'on'?
2. Is panel alive? – Warning lights 'on'? – If not:
3. Check engine fuse or circuit breaker (usually on engine) – but there may not be one.

Volvo's fuse box – 1 in use and 3 spare

This lead plugs onto the fuse in use

Small Yanmars have one but it isn't mentioned in the handbook – it's 30 amp and in the harness close to the starter motor. It's taped over and sprayed with engine paint.

Yanmar's hidden fuse

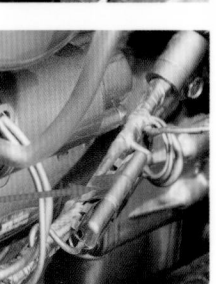

4. If warning lights go out when you operate starter, battery is flat or connections very poor.
5. If the starter solenoid is not receiving current you may be able to 'jump start' the engine. The instruments and warning lights may not then operate.
6. If the solenoid is receiving current you should be able to hear the solenoid click - if it does but the engine won't turn, then the contacts have failed or the engine/starter has seized.
7. Older engines have a spiral groove along which the gear moves to engage the flywheel. This may need to be cleaned.

JUMP STARTING

1. Make sure that nothing you are wearing can get caught in the machinery.
2. Bridge the positive (battery) terminal on the starter solenoid and the starter switch terminal on the solenoid to turn the engine.
3. This won't work if the solenoid has failed.

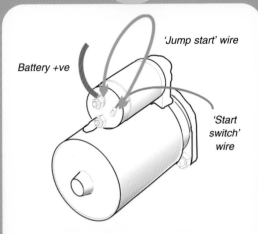

Battery +ve

'Jump start' wire

'Start switch' wire

ENGINE TURNS BUT WON'T START

TURN OFF THE SEA-COCK before continuing – turn on when engine starts.

ENGINE TURNS SLOWLY:

1. Low battery state
   – parallel batteries if possible.
2. Check battery connections – remember the battery connection to the engine block.
3. If you have decompressors, decompress the cylinders while you turn the engine with the starter so that full turning speed is achieved. Then close them (or only one if the system allows) to achieve a start.

DO NOT USE STARTING FLUID ON A DIESEL ENGINE – **DAMAGE IS LIKELY** – But you can use a hair dryer to heat the engine air intake with the air cleaner removed.

ENGINE TURNS NORMALLY:

1. Check 'stop control' is not at 'stop'.
2. If the engine is stopped electrically (no pull handle) the stop solenoid (which requires 12 volts to 'stop') may have jammed in the stop position. Some marinised engines (and generating sets) need 12 volts to run so make sure the connections are sound.
3. Check fuel 'on'.
4. Check fuel contents.
5. Check 'cold starting procedure' is correct – see engine manual.
6. Bleed low pressure fuel system – make sure it's fuel not water – see step 1 below.
7. Bleed high pressure fuel system – see step 2 below.

HOW TO BLEED

STEP 1

1. Open the engine filter's bleed screw about 2½ turns.
2. Operate the fuel lift pump lever until fuel comes out (continue until no bubbles). If the pump won't prime, blip the starter motor.
3. Close the bleed screw.

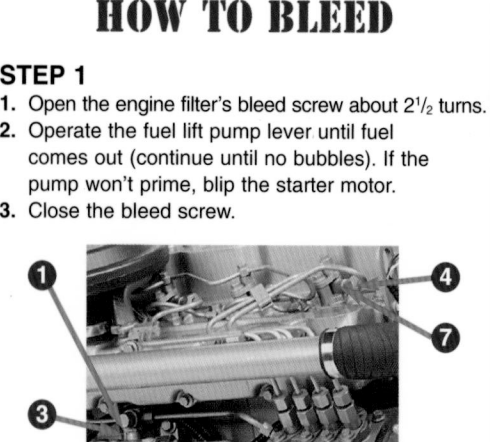

STEP 2

4. If engine doesn't start, unscrew injector unions about 2½ turns.
5. Turn off seawater inlet cock on a sailing boat.
6. Turn engine on starter motor (no more than 30 seconds, but see handbook) until fuel is seen at the loosened joints.
7. Tighten injector unions, then open sea-cock.

ENGINE WON'T STOP

Diesel engines are stopped by cutting off the fuel supply at the injection pump. Prepare for the day it fails by familiarising yourself with how it works by getting someone to operate the stop control. Look at the engine in the region of the fuel injection pump to see what moves (or makes a noise in the case of an enclosed solenoid mechanism).

If it won't stop the mechanism has failed.

ELECTRICAL STOP CONTROL – The
electrically operated stop control needs 12 volts to stop the engine.

1. If the connection has failed, connect its electrical terminal to a 12 volt supply.

2. If the solenoid has failed, you should be able to operate the lever manually.

If the solenoid has jammed and is mounted directly on the engine, removing the solenoid should shut the engine down.

Some marinised engines (and generating sets) need 12 volts to run, so if this is the case the solenoid may have seized. Remove the solenoid.

MECHANICAL STOP CONTROL
Operate the stop lever on the engine.

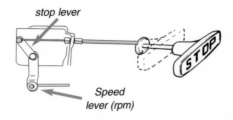

stop lever

Speed lever (rpm)

STOP

RAW WATER PUMP

– often called the sea water pump.
Practise servicing the raw water pump before you
have to change an impeller in anger at sea. (On
some engines – the smaller Yanmars in particular –
it's easier to remove the pump first.) Fitting a
Speedseal (from True Marine) will aid changing an
impeller dramatically, as no tools are required
to remove the face plate. (Sabb engines
have a diaphragm pump which
has no impeller so their owners
need to see the handbook.)

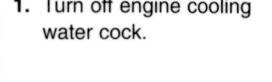

1. Turn off engine cooling
   water cock.

2. Remove the
   pump face
   plate – don't
   drop the
   screws – take
   care handling
   the gasket.

3. Remove the
   impeller. Pliers or
   slip pliers are better
   than screwdrivers –
   don't damage the
   mating surfaces.
   A large impeller will
   need a lot of 'pull' -
   special pullers are
   available but costly.

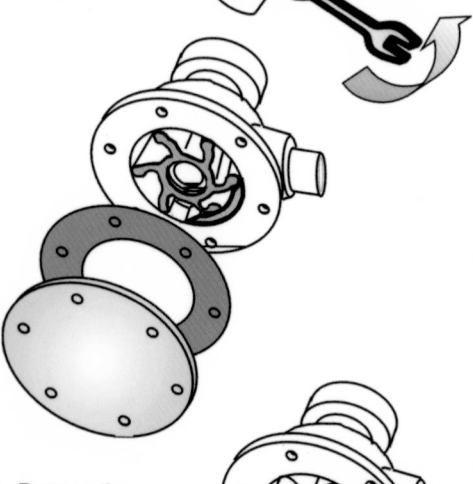

4. Examine the impeller for deterioration, especially cracked blade roots. If in doubt, renew it.

5. Check security and condition of 'cam'.

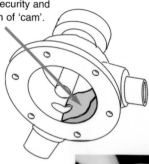

6. A cable tie or cord round the blades can be used to 'shape' the impeller to aid insertion.

7. Lubricate the impeller and refit - direction of rotation is the 'long way round from inlet to outlet'. Endeavour to get the blades pointing in the correct direction of rotation.

8. Refit the face plate using a new gasket (if necessary) – grease will help seal the gasket.

9. Insert **all** the screws before fully tightening them.

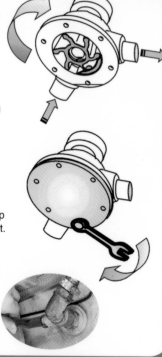

10. Open the sea-cock.

INACCESSIBLE PUMPS

Some pumps are inherently inaccessible because of the engine design, such as the Yanmar 2GM & 3GM. Some are inaccessible due to a poor engine installation.

Yanmar pumps are best removed to aid changing the impeller.

1. Undo the two retaining bolts.
2. Rotate the pump to reveal the face plate. A 'Speedseal' by True Marine aids impeller changes.
3. If possible gain access to difficult pumps by installing a hatch.

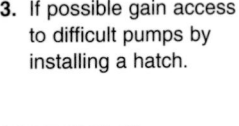

SERVICING

The best guide to servicing is the engine handbook. Although service intervals are normally governed by the number of engine hours since the last service, for the leisure boater it's more often dictated by the need to carry out a service at least annually.

COOLING SYSTEM

● **Indirectly cooled engines** need to have the coolant replaced every second year. The coolant should be made up according to the engine handbook's instructions but is often 50% water and 50% antifreeze. **Antifreeze** contains corrosion inhibitors, which are consumed over a period of time and must be replaced. Cylinder head gasket failure is common on engines which do not have this done.

● **Raw water cooled engines** often have sacrificial zinc anodes to prevent electrolytic corrosion of internal parts. These must be checked annually and replaced if more than half consumed.

The **heat exchangers** of 'indirect' cooling systems may have anodes, which need to be checked. These anodes may be found in the raw water section of the heat exchanger and may also be found in oil, gearbox and turbo intercoolers as well. The handbook should tell you. If in doubt check with the agent.

Engine anode

Oil cooler anode

Heat exchanger anode

*New anode
(but screw anode
to holder tightly)*

It's generally recommended that the **raw water pump impeller** is changed annually.

● If your engine is mounted close to the waterline in a sailing yacht, it will probably have a **syphon break**. If it has a valve (it will have no pipe leading to a drain overboard) this must be serviced annually. Failure to do this can cause water to syphon into the internal working parts of the engine.

1: Sealing washer
2: Valve diaphragm

Take the unit apart, remove all deposits and wash the valve. Volvo units must be reassembled 'upside down' as shown on the right.

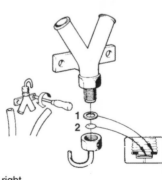

OVERHEATING is indicated by higher than normal reading of temperature gauge or sounding of alarm.

Is water coming out of the exhaust?
If not or it doesn't look sufficient, investigate:

1. Is the raw water cock open? If there is any chance that it has been turned off, check before starting.
2. Stop the engine, wait a short time and re-start. This may allow any debris to float clear of the water intake.
3. Check the raw water filter for blockage.

4. A blocked raw water intake can often be cleared by using a gas foghorn or dinghy pump blowing into the disconnected intake water pipe as close to the sea-cock as possible.

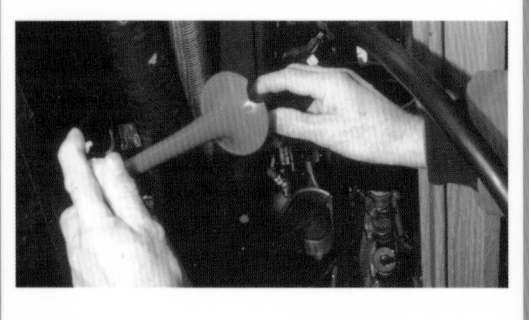

5. Raw water filter top not sealed properly.
6. Check the raw water pump. If the impeller looks intact, turn the engine over to confirm that the impeller turns and that the 'cam' remains in the correct place.

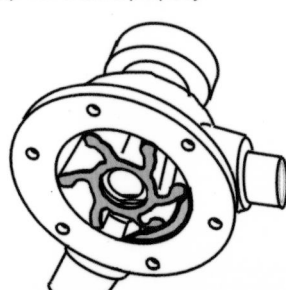

IF WATER IS COMING OUT OF THE EXHAUST

1. Old style 'bellows' type thermostats normally fail closed causing overheating. Don't run the engine without the thermostat, wedge it open instead. If the thermostat housing is 'cold' the thermostat is closed.

2. Air lock in cooling system. Some 'indirect' cooling systems need to be 'bled' to remove air if the system has been drained - see the handbook.

3. Check for soft or damaged hoses and leaking joints.

4. Blockage of tubes in heat exchanger by deposits or debris - clean the 'tubes'.

5. Blockage of internal cooling waterways – especially older raw water cooled engines.

6. Cylinder head gasket failure or cracked cylinder head.

Some engines, such as Beta, have a high operating temperature and rely on the addition of anti-freeze, as well as pressurisation, to ensure that the coolant won't boil. Because of the high setting of the overheat switch, it will not detect overheating unless the correct quantity of anti-freeze is used.

THERMOSTAT

The thermostat controls the water temperature so that the engine can run at an efficient temperature. It will normally be found in a housing and is usually easily accessible by removing a couple of bolts. Ensure the replacement is the correct type for 'direct' or 'indirect' cooling system as appropriate.

Typical thermostat housing.

Housing removed.

The thermostat.

- Correct operation of the thermostat can be checked by putting the thermostat in a pan of water on the stove. Its opening temperature is indicated on the unit and should be shown in the handbook.

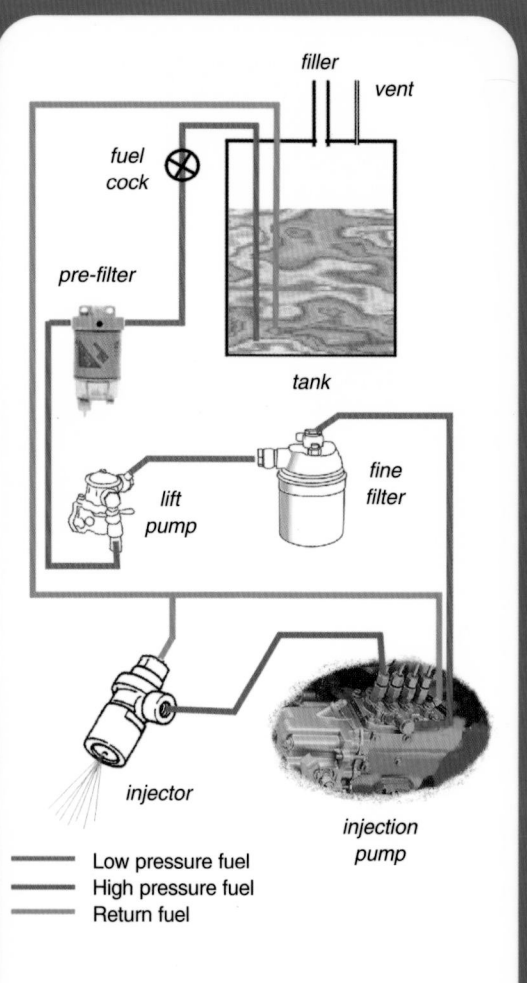

filler

vent

fuel cock

pre-filter

tank

fine filter

lift pump

injector

injection pump

——— Low pressure fuel
——— High pressure fuel
——— Return fuel

WATER IN THE FUEL SYSTEM

Severe damage to the fuel injection pump and injectors can occur if the fuel is contaminated by water.

Water can enter the tank:
- By taking aboard contaminated fuel.
- By sea or rainwater entering an improperly-sealed fuel filler.
- From condensation within the fuel tank – keep the tank full, especially during the winter.

Water in the tank can promote the growth of bacteria and fungus at the interface of the water and fuel – the dreaded 'diesel bug' – as well as corrosion in a mild steel tank.

- A water separating pre-filter (agglomerator) will stop water getting to the engine.
- Water must be removed from the tank to prevent diesel bug and corrosion.

This filter was found to be severely contaminated by 'diesel bug' when it was cut open

- Regular use of a biocide will cause resistant strains of diesel bug and should be avoided.
- Water absorbing chemicals will increase the chance of diesel bug and cause corrosion in the fuel injection pump in an infrequently used engine, as water droplets will sit in the pump for long periods of time.
- A fuel enzyme (such as Soltron) will kill the bug without the side effects of a biocide, clean the tank and improve fuel consumption. But it won't remove water – to do this you'll need to drain the tank, or syphon it out.

FINDING A FUEL LEAK

A fuel leak in the system below the level of fuel in the tank can normally be found because the surrounding area will be 'wet'. Turn off the fuel, dry the area, 'puff' with talcum powder and turn on the fuel. The leak will be shown by discoloured talcum. If the leak is in the 'suction' pipe between the tank and the fuel lift pump and above the level of fuel in the tank, air will be sucked in, rather than fuel leaked out. This is more difficult to find.

If the tank can be pressurised using a dinghy pump (with the fuel vent and filler pipes being made airtight) air or fuel will bubble out of the leak. Washing up solution can be used at the joints to trace the leak by looking for air bubbles.

SERVICING THE FUEL SYSTEM

1. The **fuel tank** should be cleaned approximately every five years, but this is seldom done.
2. Turn off the **fuel cock**.
3. Replace the filter element of the pre-filter and clean the water separating bowl. Lubricate the sealing gasket for a 'screw on' filter.

4. If fuel level in the tank is above the filter, bleed the filter by opening the bleed screw until fuel runs out. Close the bleed screw.

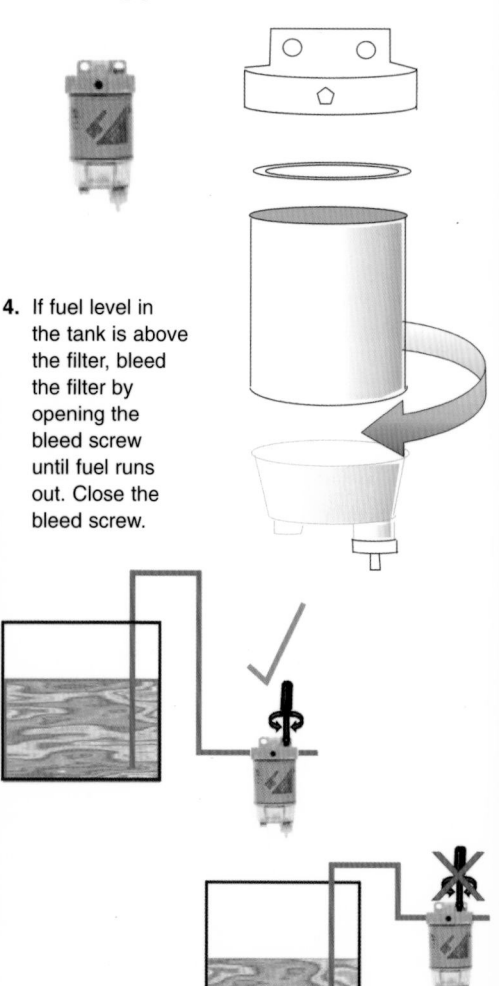

5. If fuel level is below the filter do not attempt to bleed - wait until you bleed the engine fine filter using the fuel lift pump.

FUEL LIFT PUMP

6. Most newer pumps cannot be serviced and must be replaced in case of failure.

7. Older pumps can be taken apart and replacement parts fitted. This type of pump may have an integral filter. Where a pre-filter is incorporated in the system (as it should be) this filter is unlikely to become blocked.

8. If the level of the fuel in the tank is higher than the engine the pump will be redundant and its failure (other than blockage) would pass unnoticed. Once fuel level drops below pump level, its operation is essential.

9. Clean the filter element (if fitted) of the **fuel lift pump**.

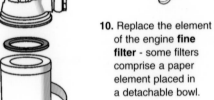

10. Replace the element of the engine **fine filter** - some filters comprise a paper element placed in a detachable bowl.

11. Lubricate the sealing gasket for a 'screw on' filter.

12. **Bleed** the system.

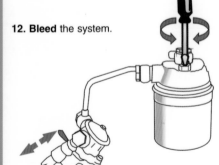

13. Service or replace the fuel injectors if symptoms dictate – light grey to blue smoke all the time and difficult starting.

- Change the oil 'on schedule' or more often if you have a lot of 'stop/start motoring'.
- Use the grade and viscosity of oil as indicated in the handbook – in the case of older engines the specified grade may not be obtainable so use the nearest. In many sailing yachts, with minimal use, engines will often never become fully 'run in', even by the time the engine dies a natural death. In these situations, the use of 'super oils' often compounds the problem.

ENGINE OIL CHANGE

1. Most people recommend that the engine should be run up to warm the oil, making its extraction easier.
2. Let the engine stand for 10 minutes to allow the oil to settle.
3. Remove the dipstick and insert the tube of the oil extraction pump, trying to get it as far to the bottom as possible.
4. Pump out the oil.

5. Pour the required quantity of new oil into the oil filler. (This won't be contaminated by the dirty oil still contained in the old filter, but will give time for the oil to reach the sump.)

6. Using a suitable wrench, remove the old oil filter, trying to contain the spilled oil.

7. Fit a new filter, first lubricating the oil seal, and tighten as indicated on the instructions printed on the filter.

8. Check the oil level. (As the filter doesn't yet contain any oil it may over-read.)

9. Run the engine for a couple of minutes to check for leaks. If you have a mechanical 'stop' control, keep it pulled until the oil pressure light goes out, to allow oil pressure to build before the engine starts.

10. Wait 10 minutes, check the oil level and top up if necessary.

GEARBOX OIL CHANGE

- Oil normally has to be removed by a pump, through the dipstick hole.
- For saildrive legs and out-drives, the boat will have to be out of the water to change the oil, as oil is drained from the bottom of the leg. (Remove oil filler cap first.)

DRIVE BELTS

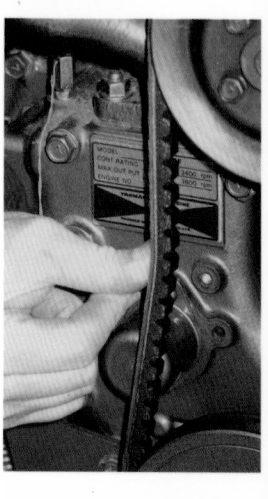

1. Drive belts must be maintained at the correct tension.
2. The normal 'rule of thumb' is to press the belt and check the deflection is about 5 mm.
3. OR, If you twist the belt between thumb and forefinger it should turn approximately 90 degrees.
4. After fitting a new belt, tension will need to be adjusted.

DRIVE BELT ADJUSTMENT

Generally, with the attachment bolts slackened, the alternator is levered away from the engine (A) while tightening the adjustment bolt (B). Re-check after 5 hours.

ALTERNATOR

- Check all connections for tightness and corrosion.
- DO NOT switch off the battery switch while the engine is running because destruction of the alternator's diodes may occur.

On some engines, switching off the 'ignition switch' may have the same effect.